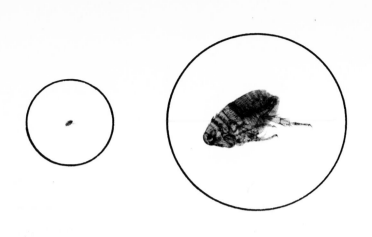

ZOOMING IN

ZOOMING IN
photographic discoveries under the microscope

Text by Barbara J. Wolberg   *Photographs by Dr. Lewis R. Wolberg*

HARCOURT BRACE JOVANOVICH  New York and London

For MICHAEL and DAVID

Printed in the United States of America

First edition

B C D E F G H I J K

Library of Congress Cataloging in Publication Data

Wolberg, Barbara J
Zooming in.

SUMMARY: Introduces in text and illustrations the
process and techniques of photographing a variety of
items, including rocks, metals, chemicals, food, plants,
and animal tissue, through a microscope.
Bibliography: p.
1. Photomicrography—Juvenile literature. 2. Microscope and microscopy—Juvenile literature. [1. Microscope and microscopy. 2. Photomicrography]
I. Wolberg, Lewis R., illus. II. Title.
QH251.W58   778.3'1   73-18631
ISBN 0-15-299970-1

Book design: Robin Fox

Contents

WE HAVE AVAILABLE to us almost the entire world of objects for observation under a microscope and photographing for our education and enjoyment—rocks, metals, chemicals, textiles, household items, food, wood, plants, insects, and animals' tissues. Everything the eye sees through the microscope we can photograph with a camera. This process is called *photomicrography*, and the pictures taken are called *photomicrographs*.

In this book representative photographs of this hidden world are presented for the simple pleasure of viewing or as a guide for producing your own photomicrographs. Some of these samples were made from slides prepared through a complex but easily learned technique that is fully detailed in the appendix by Dr. Lewis R. Wolberg, Clinical Professor of Psychiatry, New York University Medical School.

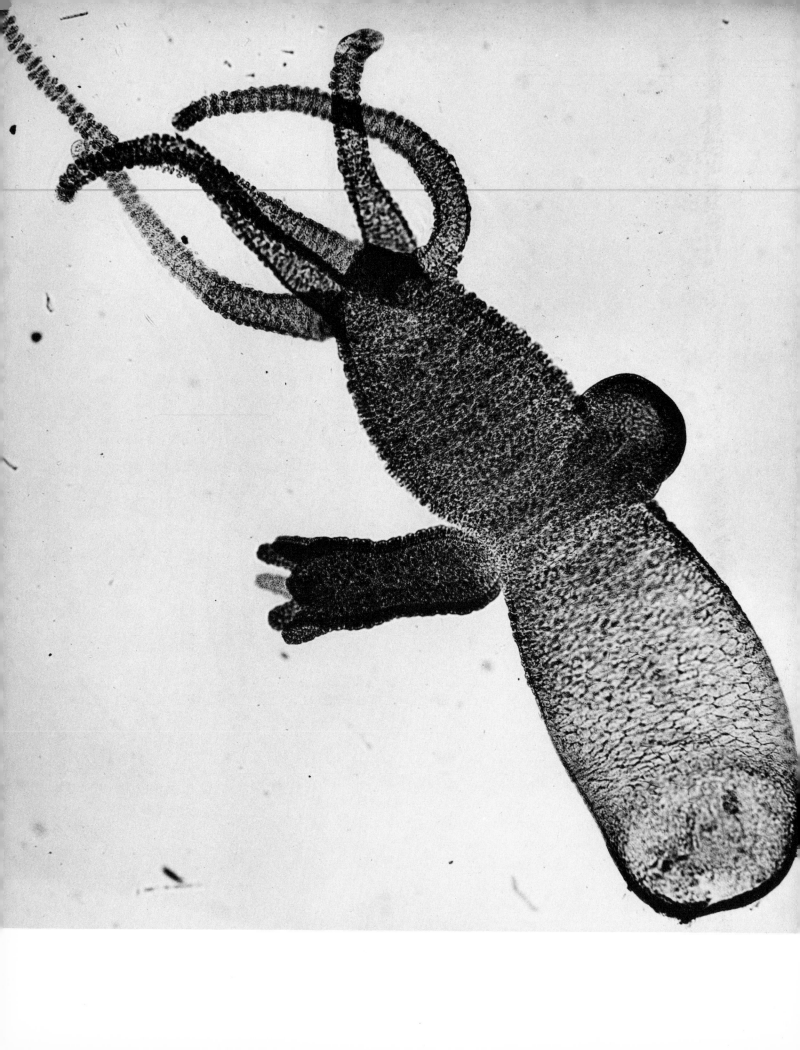

1. A Tiny Aquarium

IMAGINE YOU ARE LOOKING through the lens of the microscope at a single drop of pond water. At first, the water seems perfectly clear, but as you focus the lens, tiny specks appear; they grow larger and larger and seem to be alive—little creatures of different shapes moving in every direction.

Many of the shapes are curious little animals. A hydra reaches its tentacles out in search of food, sending some water fleas scurrying out of its way. An amoeba swims by slowly, while some larger oval-shaped creatures known as paramecia glide past, munching on tiny plants called bacteria. The most unusual animals are clusters of Obelia that look like a bouquet of flowers.

In the drop of water there are also many beautiful plants known as algae. The loveliest are the diatoms, which resemble wonderfully decorated beads. Others twirl around each other in tangled, twisted masses or live together in groups called colonies.

Although these plants and animals look strange and unfamiliar, they are not very different from larger organisms. Their bodies are made up of individual units called cells, each of which contains a series of substances and structures that together make up protoplasm, the basic material of life.

◀ Coelenterates are "hollow-tube" animals that vary in size from the almost microscopic hydra to the giant jellyfish. The mouth—which is the only opening in these animals—is surrounded by a ring of tentacles and is used not only to take in food but also to expel solid wastes. The hydra uses its tentacles to capture and paralyze its prey and then to push its captive through its mouth into the digestive cavity. One way in which the hydra reproduces is to produce buds from its own body that separate and form new animals identical to the parents. You can see a tiny hydra growing out along one side of this hydra's body, while the bump on the other side is another bud beginning to form. *(160x)*

Little crustaceans sometimes called water fleas belong to the same family of animals as the insects. The larger creature known as Daphnia is a common species and is especially interesting because you can observe its internal organs through its transparent outer skeleton. By studying Daphnia, scientists first discovered that white blood cells eat bacteria, thus keeping the blood clean. Many diatoms appear as tiny circles, strands, and boat-shaped objects. Another species of water flea can be seen at the top. All these plants and animals make up part of a community called plankton, the main food supply for all kinds of marine life. *(75x)*

Two paramecia show why these one-celled animals have been described ▶ as "slipper animalcules." A special silver stain was used to bring out the way the skin is dotted with rows of granules. These form the bases of little projections called cilia, short hairlike structures like eyelashes. These hairs beat in unison, propelling the paramecium through the water at great speed. The cilia also create currents that sweep food particles into the mouth or oral groove, which you can see on the paramecium at the top. Paramecia live on bacteria, and one animal can eat up to five thousand bacteria a day. *(900x)*

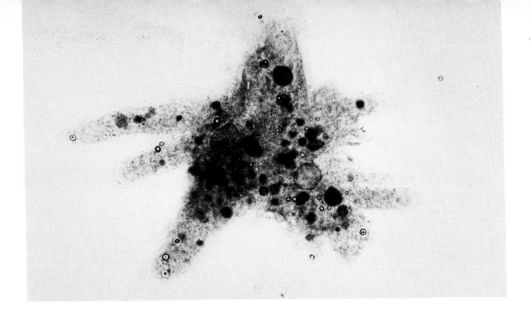

The amoeba pictured clearly demonstrates how this one-celled creature gets about. Since it has no special organs for locomotion, it moves by thrusting out extensions of its body like false feet. Inside the body of the amoeba many different-sized bubbles and grains can be seen, some of which are fat droplets, food particles, waste materials, or digestive and water vacuoles. All these, along with the nucleus—which is the largest irregular oval and barely visible—make up the protoplasm of this animal. *(550x)*

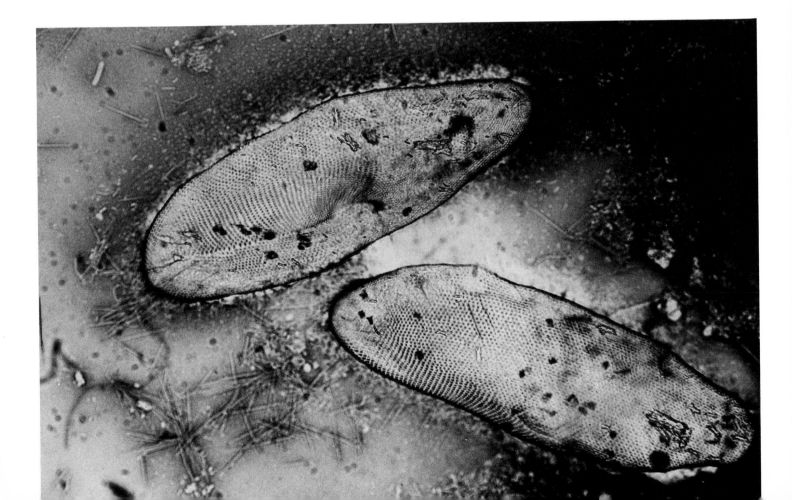

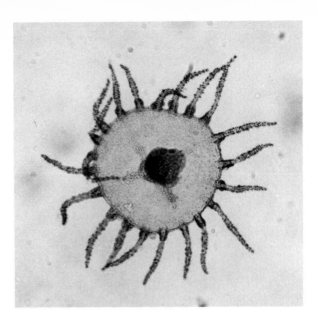

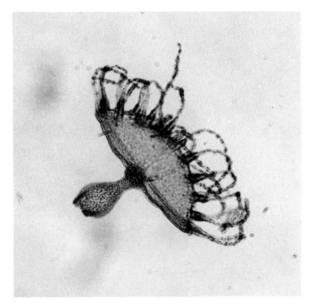

Groups of water animals, Obelia, live together in a colony. They have polyps that look like tulips attached to stalks, which give the entire colony the appearance of a plant. Each polyp consists of a transparent cup of stiff, horny material into which the animal can withdraw when alarmed. Each has a mouth and tentacles to capture food. The stalks develop buds, most of which gradually develop into new polyps. Others develop into medusas—bell-shaped or umbrella-shaped bodies—which separate from the stalk and swim away. After a period the medusas release eggs and sperm that after fertilization develop into polyps. These in turn start a new colony. *(Left, 160x; top, 300x; bottom, 300x)*

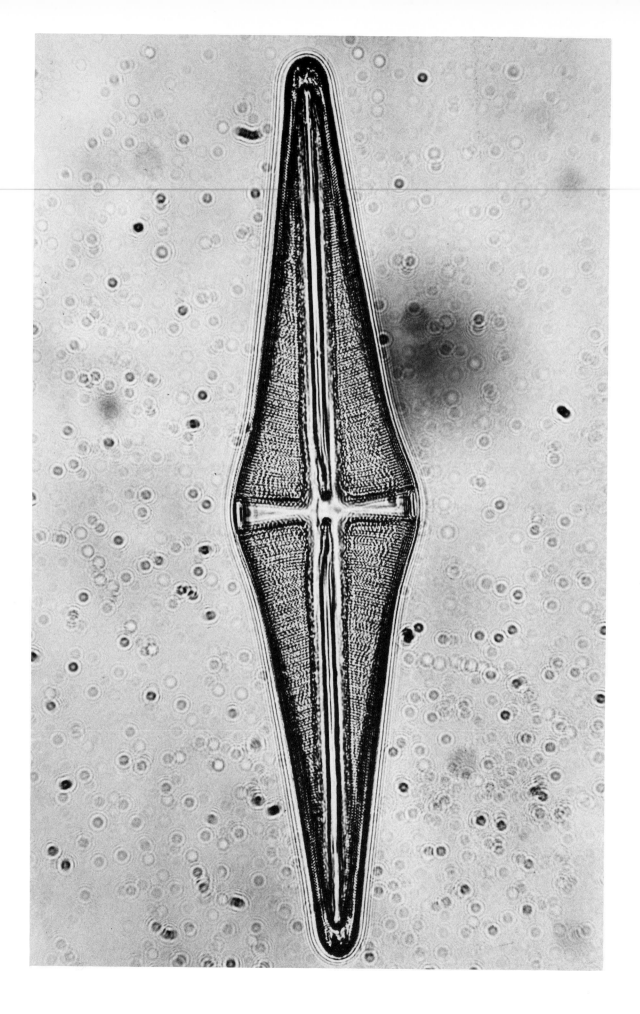

◄ Diatoms are one-celled plants that occur in over five thousand different species. They are found in countless numbers in the water. This single diatom shell shows the cell consisting of two similar valves running the length of the plant, connected to a band or girdle running through the middle. The valves are decorated with fine lines of dots and circles. Most diatoms reproduce by splitting into two halves where the valves join. *(1,450x)*

Colonies of Zygnema are cells of algae strung together like beads on a necklace. The nucleus of the cell is seen as a large, dark circle. Some of the nuclei have lost their shape and are in the process of dividing. Chemical messengers constantly pass in and out of the nucleus into the body of the cell, directing the many chemical activities that take place inside that cell. *(750x)*

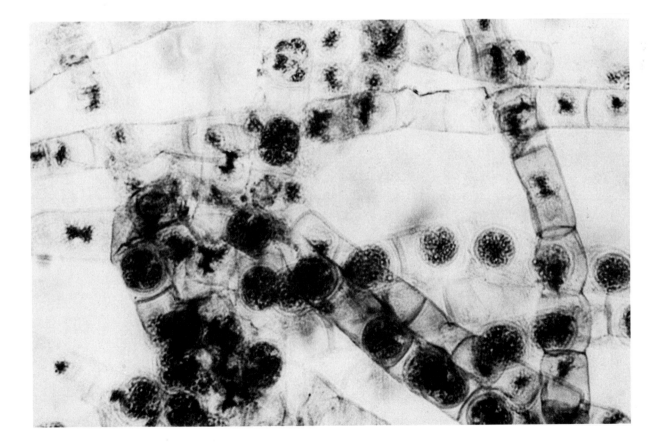

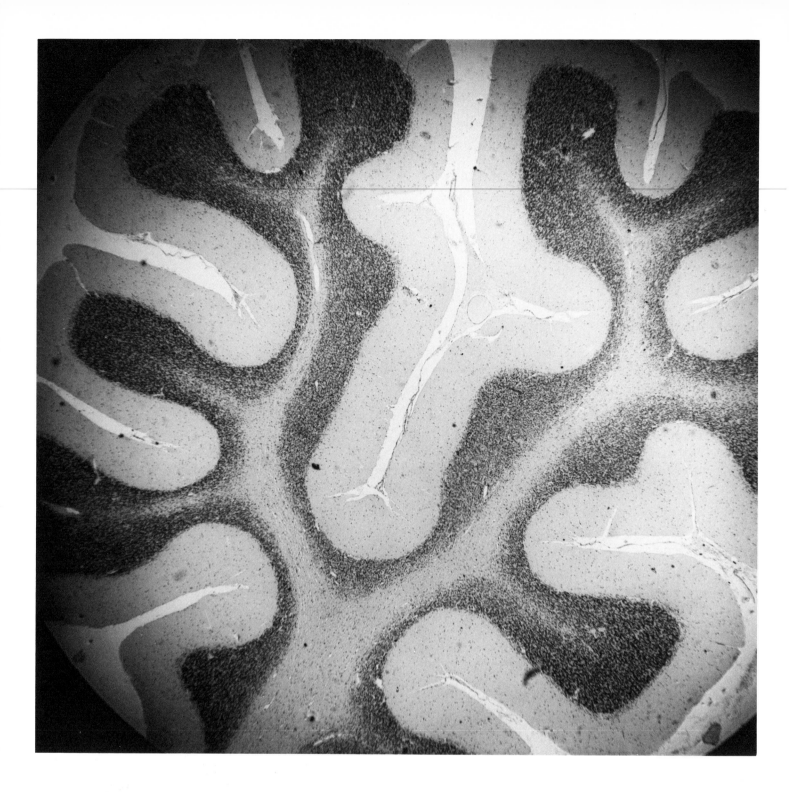

The cerebellum is the arranging and coordinating part of the brain. You rely on the cerebellum for moving and balancing.

The large, irregular shapes in the cerebellum seen here are folded bundles of nerves, which show up as fingers of black nuclei surrounded by grayish areas of fibers. The white streaks are spaces between the nerve bundles. *(65x)*

2. Cells

THE AMOEBA AND THE DIATOM are able to live alone as single cells, but the majority of cells must live in groups joined together with other cells. These groups link together to form the tissues and organs of all plants and animals.

With the microscope you can see some of the groups of cells that make up the human body. The nerve cells of the brain form patterns that swirl and dip and curve. The muscles and the blood vessels of the lips, ear, and intestines make sharp contrasts in rich shades of black and white. The differences you see are matched by the differences in function, because nature has designed and arranged these groups of cells according to the jobs they must do. Some cells are responsible for digestion, some for movement, others for respiration or reproduction.

A fertilized egg grows into an embryo and then into a baby. As the baby grows into an adult, new cells are added at the same time that worn-out cells are destroyed. For example, the red blood cells are replaced on a regular basis. The skin cells need to be replaced every five days or so; the lining of the stomach and the intestines even more frequently. Other body cells, like those of the liver and the kidneys, replace themselves very slowly.

Although cells are shaped differently and perform diverse functions, each contains the same important parts and chemicals. The most vital part is the nucleus, which is the control center for the cell. Inside the nucleus is a material called DNA, and it is by means of this remarkable chemical that the nucleus directs all the activities of the cell and keeps the whole organism alive and well.

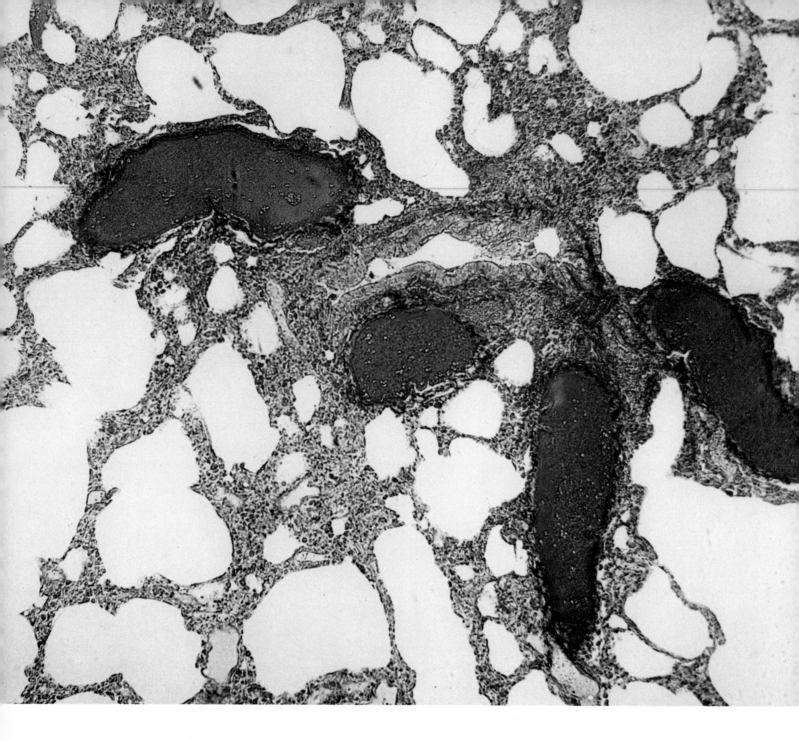

▲ When you breathe in air through the nose and mouth, it goes down the trachea and then to the larger bronchial tubes that lead into the lungs. In this section of a human lung you can see many open spaces. These are sacs made up of cells lined by a thin membrane under which the blood corpuscles circulate to collect oxygen and release carbon dioxide. The blood corpuscles enter blood vessels (like the four finger-like, elongated circles in the photograph) and carry oxygen to all parts of the body. *(200x)*

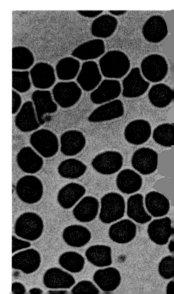

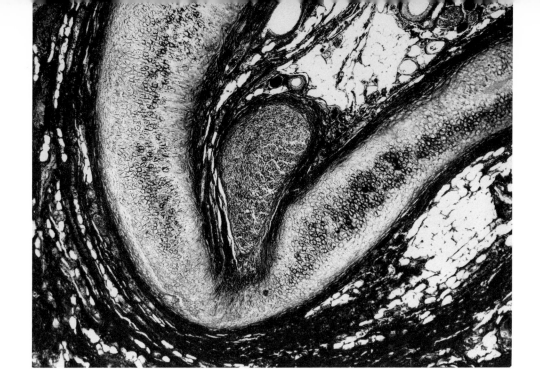

The V-shaped gray area here with tiny cells is a section of the cartilage of the outer human ear. The cartilage is surrounded by white spaces and black streaks, which are the blood vessels and softer connective tissue.

The human ear has three parts. The outer ear collects the sound, sending it through the middle ear to the nerves in the inner ear, which go on to the brain. *(50x)*

There are several kinds of cells shown in this smear of human blood. The small, round gray discs are the red blood cells. Scattered among these are a few white blood cells, which appear as colorless bodies containing black nuclei in different shapes. There are several different kinds of white blood cells, some of which serve to fight off infections. The red blood cells move oxygen into the body from the lungs. *(1,200x)*

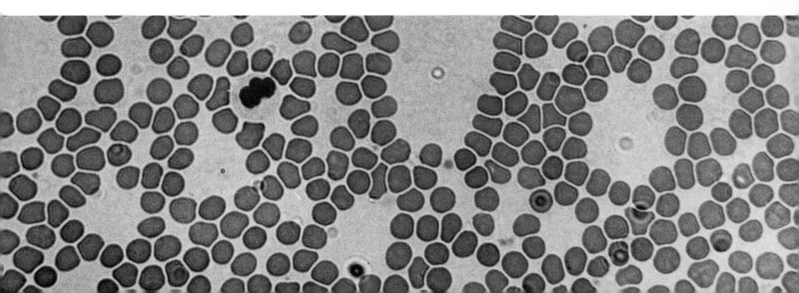

This is a cross section of the outer layer of the tongue, showing ▶
the taste buds (dark part) and the shaggy muscle tissues (light gray
part).

    The top of the tongue has many small bumps on it called papillae.
These contain nerves, which make it possible to taste food. Each part
of the tongue is sensitive to a different taste; at the tip, you taste sweet
and salty; on the outer edges, sour; and at the back, bitter. *(375x)*

In this section of the inside of a human lip, the black bundles are the
muscles, and the grayish streaks the blood vessels and connective tissue.
The lips contain a rich supply of blood vessels and muscles. *(300x)*

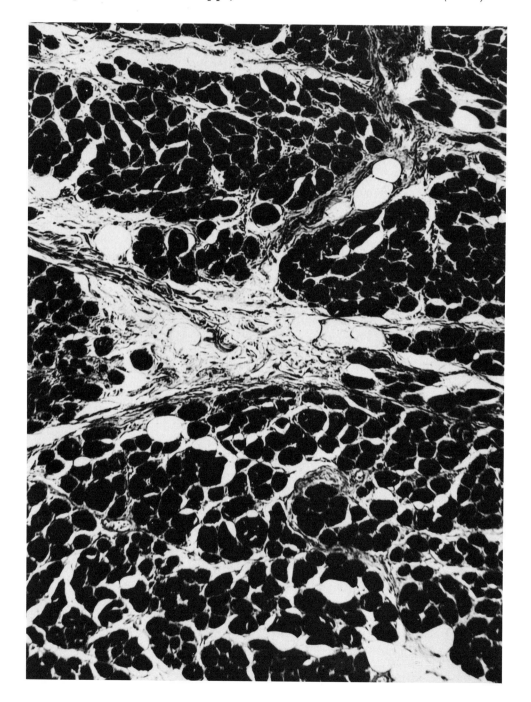

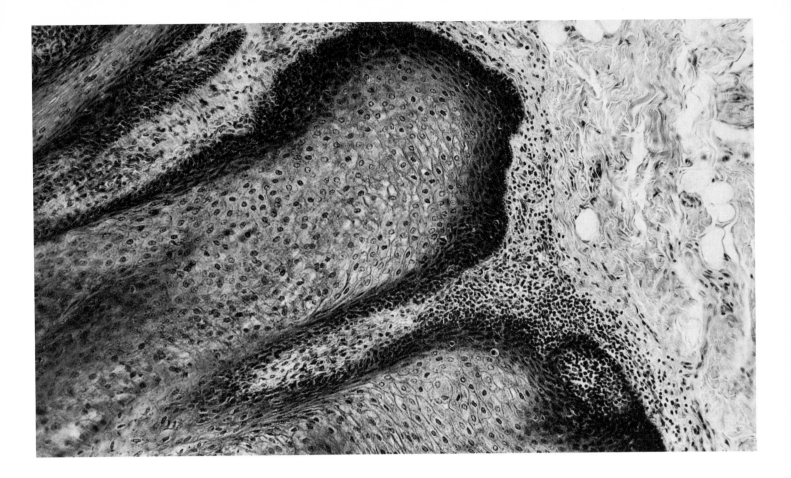

This cross section of human bone shows the calcified outer cells as black spots, the fibrous tissue as gray areas, and the finger-like marrow areas as black streaks.

Human bones are proteins that contain minerals. At birth bones are soft, but as they grow, minerals like calcium and phosphorus are deposited in them, making them hard. Bones demonstrate they are alive by mending together when they are broken. *(1,200x)*

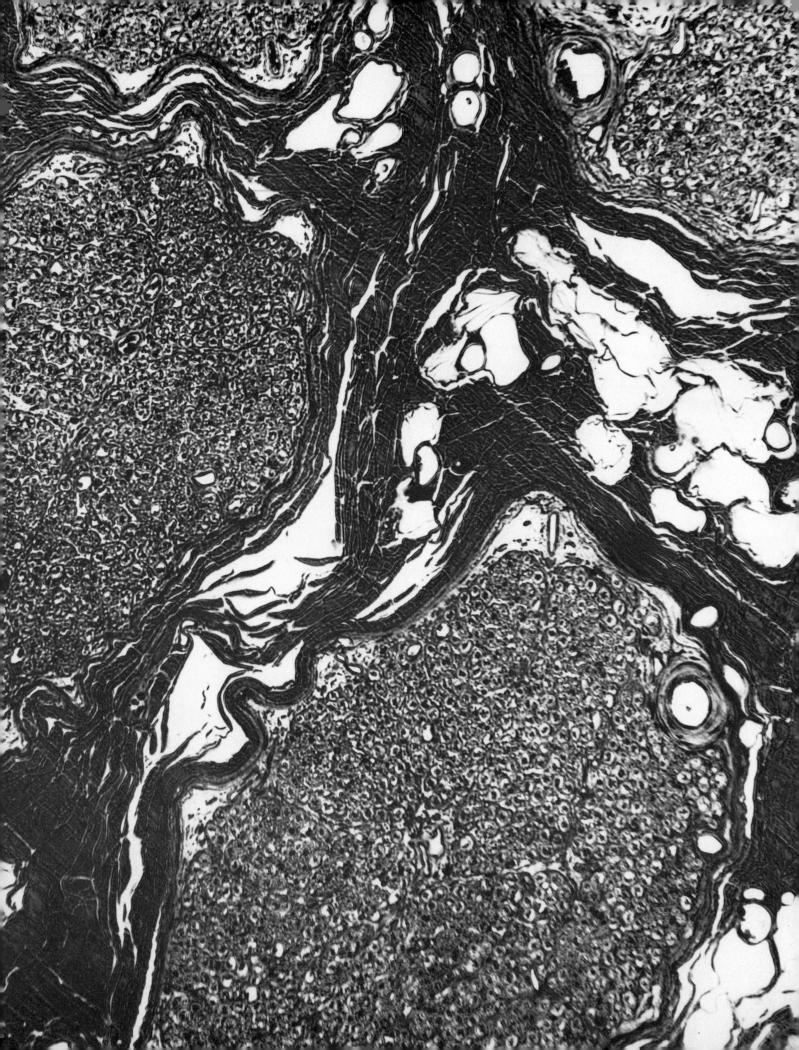

◀ This is a cross section of the sciatic nerve showing several bundles of nerve fibers (gray parts) separated by curving strands of connective tissue (black parts) with spaces, some of which are blood vessels lined by cells.

  The sciatic nerve is one of the largest nerves in the human body. It starts around the hip and travels down the thigh and leg. *(700x)*

  Inside the stomach, enzymes break up the food so it can be absorbed. Carbohydrates and fats are reduced to glucose—a sugar—and proteins are digested into chemicals called amino acids.

  This part of a human stomach shows the gastric glands as black grape-like clusters. These secrete the fluids that help digest food. The muscle fibers appear as black strands, the blood vessels and connective tissue as the grayish parts. *(415x)*

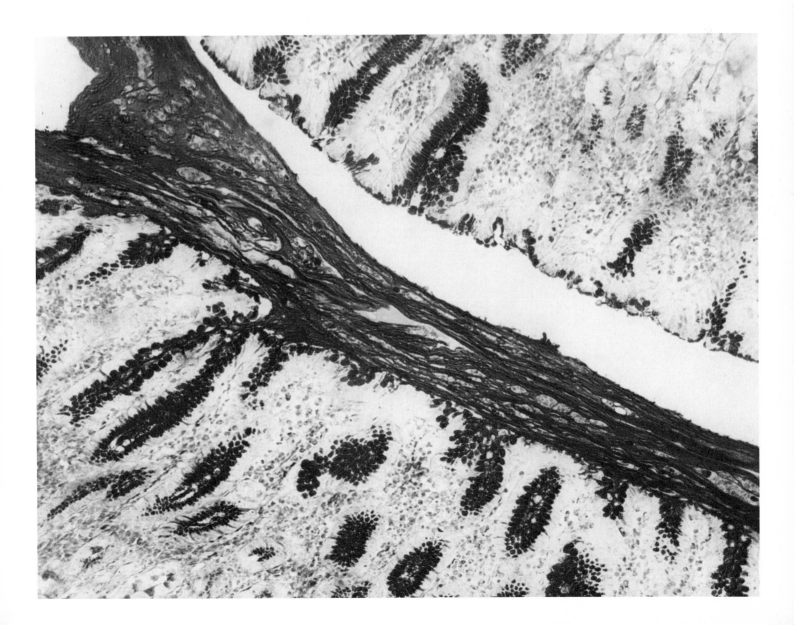

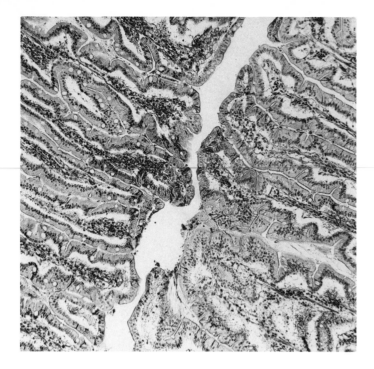
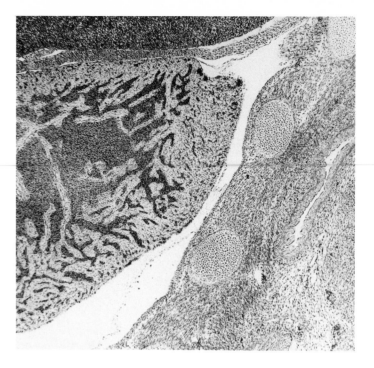

This cross section of part of the human small intestine shows the wide white space where food enters running through the center. The finger-like projections of cells digest the food, absorb it, and transfer it to the blood. Blood vessels (small white circles) are scattered throughout. Digested food enters the bloodstream through the walls of the intestines and is then carried to the cells of the body, where it is used to supply energy or to build new body cells. *(165x)*

This group of cells from the embryo of a mouse will arrange themselves into groupings that will become the organs and tissues of the body. It is the DNA in the nucleus of each cell that determines what it will become. The bones collect in the center, and the cells that will become the muscles, blood vessels, and connective tissue gather around the bones. Other cells gradually form glands and organs like the lungs and heart. *(100x)*

This section of developing tissue demonstrates how individual cells ▶ multiply. Inside each cell circle is a smaller circle, the nucleus, distinguished by its granular texture. When a cell reproduces, it grows until it doubles in size; then it divides into two. Before the cell divides, the nucleus divides. When the nucleus divides, black clumps called chromosomes appear. After the cell divides, it stays together with the parent cell, and the tissue to which they belong grows. *(3,000x)*

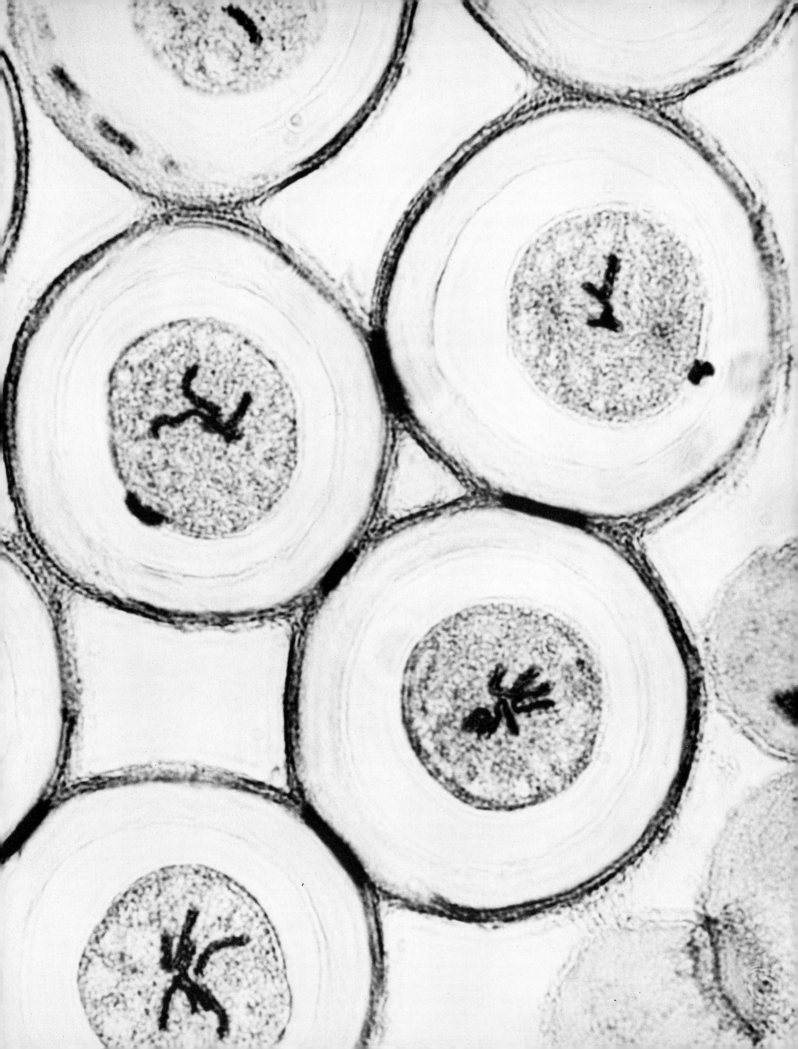

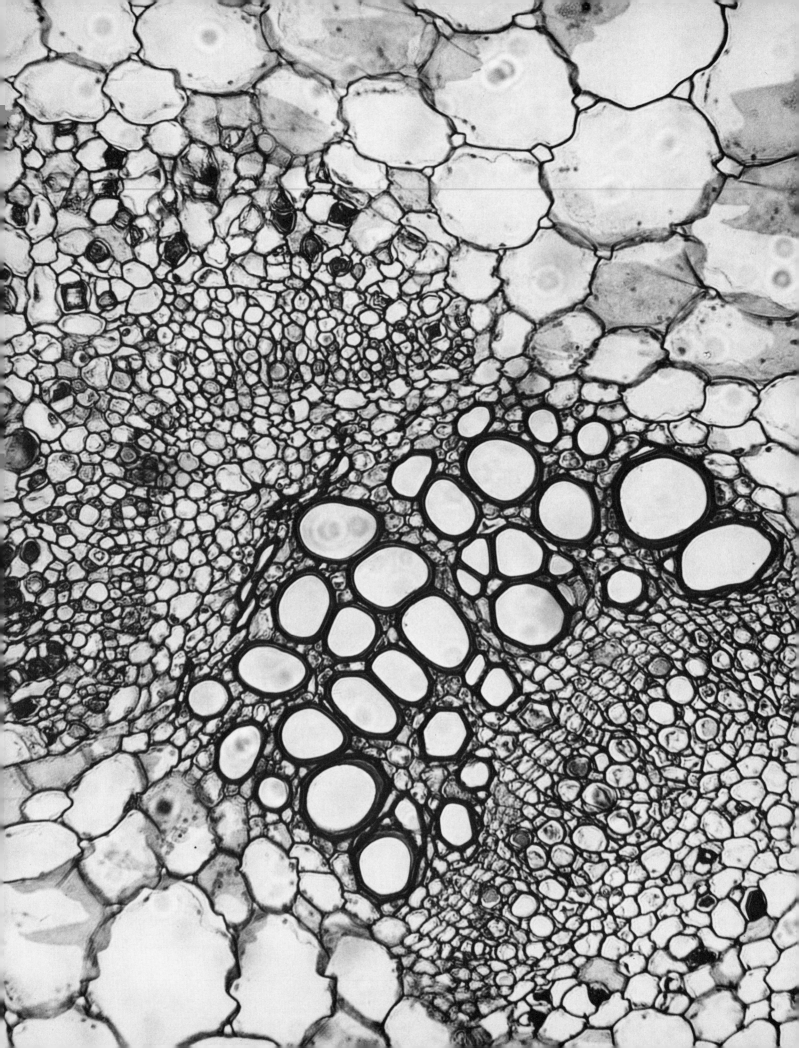

3. Food

UNDER THE LENS of the microscope a drop of chicken soup becomes unexpectedly beautiful. An onion or potato is seen as a cluster of cells with many patterns. A tiny grain of salt or sugar becomes a giant crystal formation. Food chemicals called amino acids look like splashes of paint or exquisite pen-and-ink drawings. Each amino acid has its own unique arrangement.

Carbohydrates, fats, and proteins are the foods of all living things, whether plant or animal. They are used first as organic fuel to supply energy in the processes of life and then as building material for new cells that make up the organism.

Green plants get food by manufacturing it in their leaves, using the energy from the sun. They combine carbon dioxide from the air with water from the soil to make sugar and starch. Afterwards the plant combines these substances with the minerals it has taken in from the soil to produce fats, proteins, and vitamins.

Animals must get their food from a source outside themselves. Most animals are selective about the kind of food they eat. Some eat only plants, while others eat only animals. Still others, like man, eat both plants and animals.

◀ This part of a cucumber shows large cells with thick walls that act as a structural support and also serve as a conductor of fluids. The other cells store food. All green plants accumulate starch, protein, and fat in fruits, leaves, stems, seeds, and roots. When these are consumed by man, they contribute a valuable portion of his total nutrition. *(700x)*

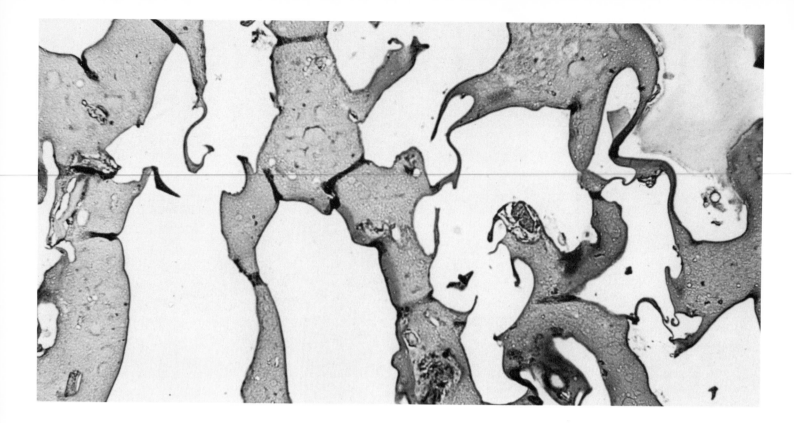

Pieces of meat and bits of vegetables, when cooked together in chicken soup, release fat globules that appear as droplets on the larger shreds of food. During digestion these fats are used to supply energy. Excess fats are stored in tissues under the skin and in and around some organs in the form of body fat. They serve as a source of reserve energy. *(200x)*

Inside the irregular-shaped cells of the potato are special storage cells that contain starch granules. These appear as little white circles and ovals crossed with delicate black markings. *(300x)*

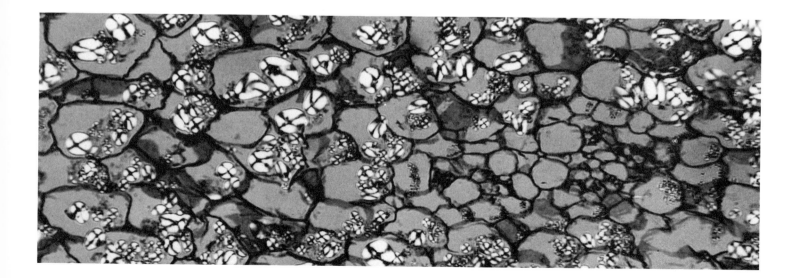

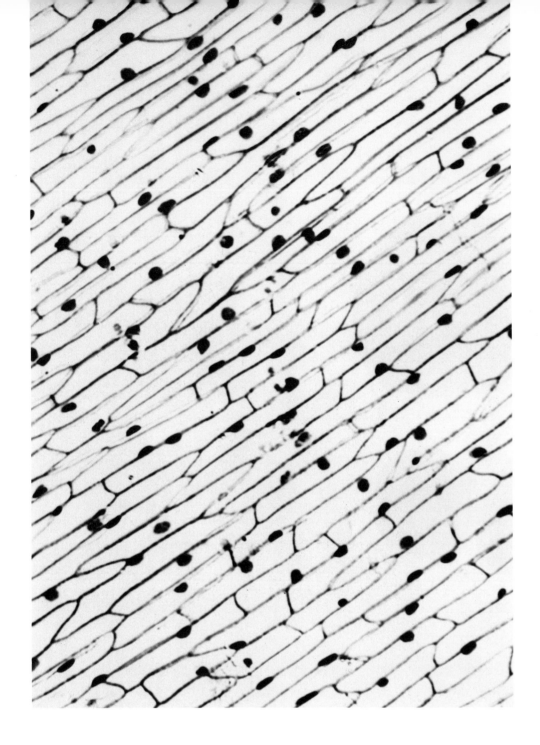

A section of the skin of an onion shows a fine network of elongated cells closely cemented together. Inside each cell, lying near the cell wall, is a black dot; this is the nucleus. The cell walls of plants are made of a material called cellulose. The cellulose cannot be digested by humans, but is used by the body as roughage to stimulate the muscle action that keeps food moving along the entire digestive tract. *(200x)*

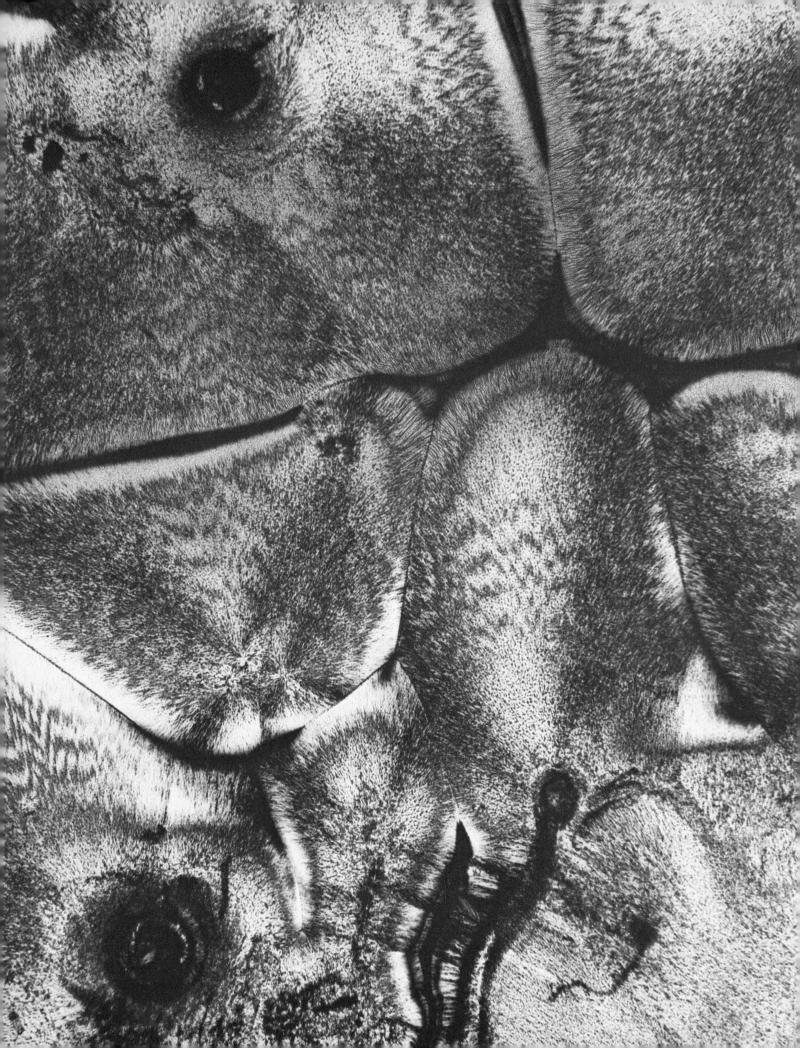

This beautiful spray reveals the crystal structure of Vitamin C. Man needs Vitamin C in his diet, while other animals and plants are able to manufacture it in their own bodies. Vitamins help in keeping the body healthy by preventing certain diseases. *(320x)*

◀ The proteins in all forms of plant and animal life are constructed from the same basic set of twenty amino acids of which arginine is one. In man, arginine is considered one of the essential amino acids and operates in the liver to manufacture urea. *(320x)*

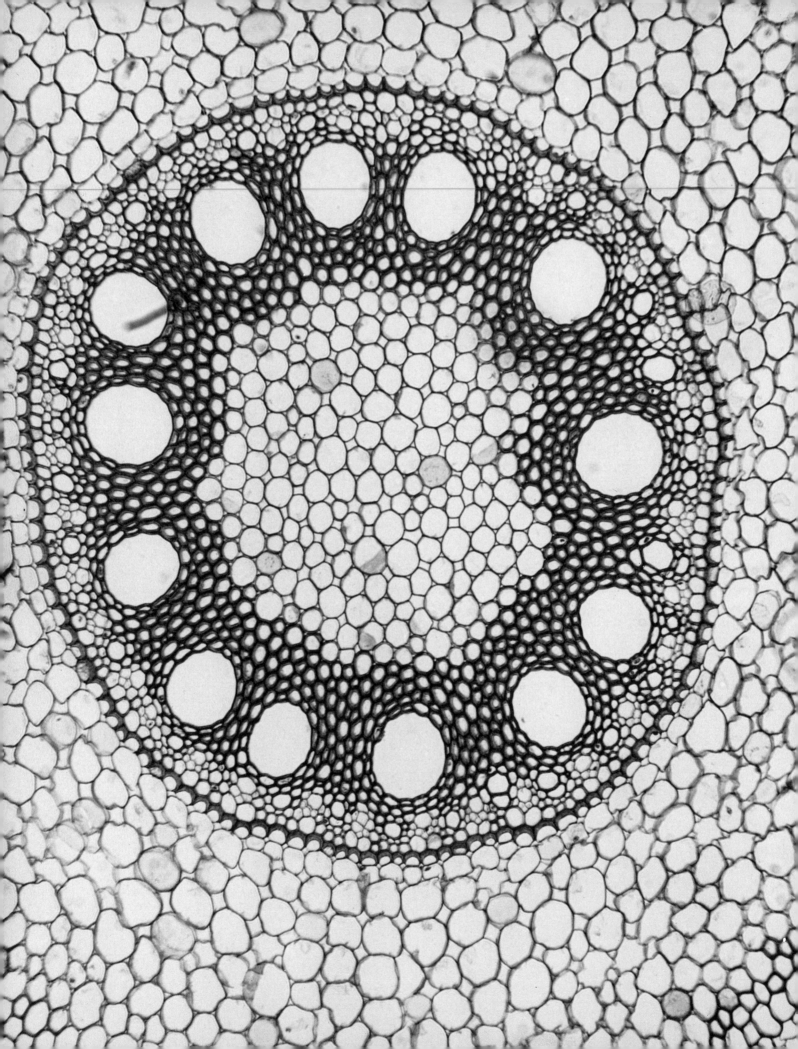

4. Plants

PLANTS ARE SOME of nature's most exquisite creations. They can put on a splendid show in almost every season. But there are parts of plants that can be seen only under magnification that are just as lovely as the whole ones. The special structure of a seed that aids in its dispersal becomes easily visible. A comparison of leaves, stems, and roots from various plants reveals minute but complex differences. Splinters of wood are fascinating because every kind of wood, no matter how small the piece, has a particular pattern formed by big and little spaces of different sizes locked together like stitches in elegant needlework.

Most plants are green, but there is a curious group of plants called fungi that are not. True fungi are masses of tiny threads that mat together and grow into odd shapes like mushrooms, molds, and mildews. Fungi play an important role in nature as decomposers. They cause the decay of plant and animal remains into substances such as water, carbon dioxide, and mineral salts that return to the air and soil to be used again in new plant growth.

◄ Most plants depend on roots for support, for carrying soil, water, and mineral salts into stems and leaves, and for storing food. In this cross section of a corn root, the intricate conducting system can be seen. The largest circles are the ducts through which water and minerals are taken into the plant from the soil and carried upward to the leaf, where they are changed into food. The smaller, darker circles surrounding the larger ones are ducts that transport the food materials from the leaves downward to the roots for storage. *(500x)*

This seed has a long beak with a crown of "feathers" that keeps it floating through the air like a kite. Each seed is an embryo, complete with stem, root, leaves, and enough food to support its growth for a few weeks in the spring while it germinates. *(160x)*

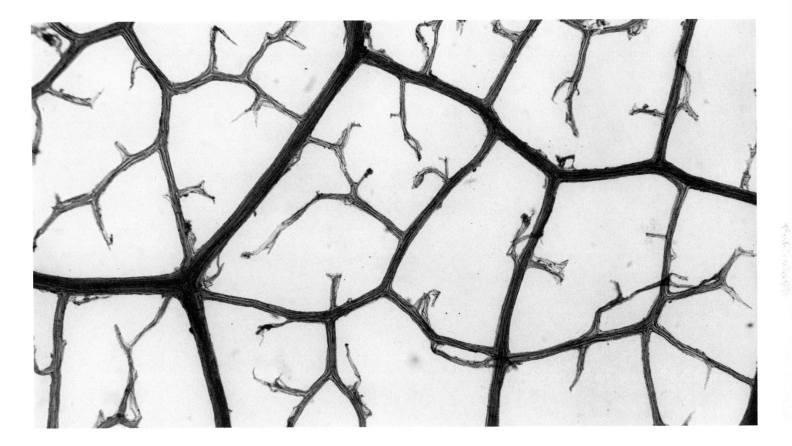

Through all broad leaves runs a network of veins that branch and crisscross and perform several functions. These veins act as a supporting skeleton for the leaf. They also contain conducting tubes that connect with the tubes in the stem, forming a continuous system of pipes through which water, food, and minerals are transmitted. *(125x)*

◀ Unlike animals, green plants have special cells that act as storage chambers for food. Inside these storage cells of the lily, accumulations of starch can be seen as clusters of white grains. This material is used by the lily as food during the winter before a new crop of leaves comes out in the spring. *(500x)*

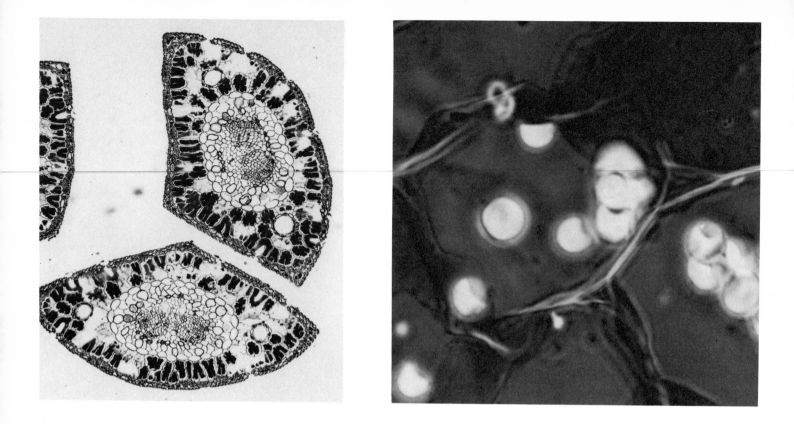

The leaves of evergreens are often shaped like needles, but they do the same work as broad leaves. In this cross section of two pine needles, the vein appears in the center as an accumulation of irregular circles surrounded by dark cells that contain chlorophyll. Chlorophyll is the chemical that enables the plant to make its own food. The two large white circles, one on each end of the central vein, are the ducts that carry resin, a substance released to seal wounds when the plant is injured. *(140x)*

These flat cells come from a plant called the liverwort. Liverworts are no more than sheets of cells a few layers thick. They form green patches on wood, rocks, and soil in moist and shady places where most other plants cannot live. Inside each cell are the starch granules, which appear as white discs. The liverwort was the first land plant to emerge from the ancient seas and is the ancestor of the mosses, ferns, and leafy plants. *(3,000x)*

Attached to these slivers of fern leaves are delicate sacs called sorti, ▶
which contain bundles of little bodies called spores. Flowering plants reproduce by means of seeds, but plants that do not flower reproduce by spores. When the spores are ripe, the sac breaks open, casting the spores into the air. *(200x)*

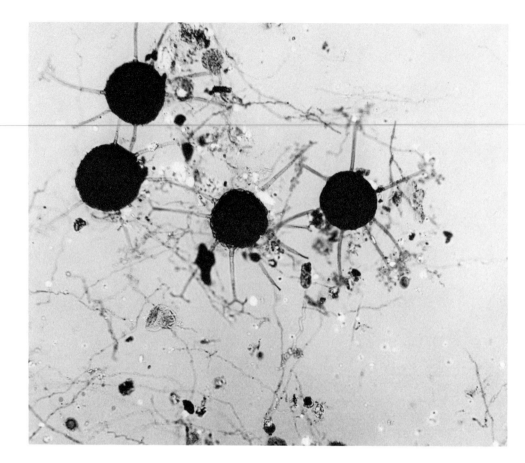

In some fungi, such as the ones in this photograph, the spores are developed in special spider-like bodies. Spores are present everywhere—in the dust, in the air, in the soil. When they have the right conditions of moisture, food, and temperature, they will grow and form new fungi. Fungi live on the bodies of either live or dead plants and animals. *(315x)*

In this cross section of white oak, the large gray circles embedded in ▶ streaks of fibrous supporting tissue are open conducting ducts. The horizontal band near the middle of the photograph is a growth ring. *(300x)*

A longitudinal section of hard maple shows bundles of small light ▶ and dark circles. These are the vascular tubes of the conducting system. They are surrounded by lines of tough fibrous tissue, which help support the tree. *(1,200x)*

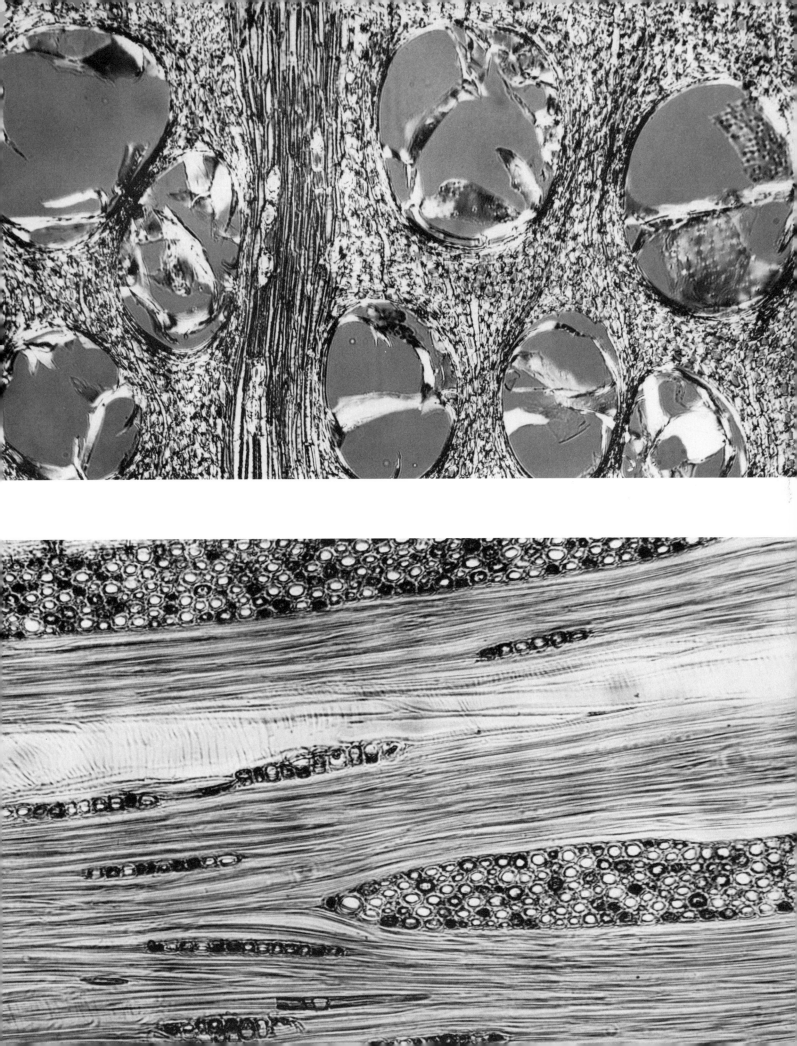

5. Animals

By attaching lenses of different strengths to the microscope, you can make more highly magnified images and observe many species of animals that are nearly or completely invisible to the unaided eye. Tiny insects can look like monsters under magnification. Minute one-celled creatures from the sea appear as giant shells that one might find along a sandy shore, yet each is smaller than a grain of sand. But every animal, no matter how small and inconspicuous, is part of a complex set of interrelationships with other animals and plants that create order in their environment.

Whole animals are by no means the only subjects that make interesting viewing. A little dust from a butterfly's wing resolves itself into beautifully formed scales. A fragment clipped from a feather, the wing of an insect, or the hide of a mammal reveals an intricate construction. Fragile bits of sponge skeleton look like silver needles or stars that make up a delicate natural maze, and parts taken from various marine animals appear as lovely abstract designs.

◀ Butterfly wings are covered with regular scales that have delicate markings on them and overlap one another like shingles on a roof. This overlapping prevents air from getting through while the wings are in use. *(1,200x)*

This fragment of down feather shows the strong central shaft out of which grows fibers known as barbs. Birds have different kinds of feathers on various parts of their bodies. Down feathers are the birds' soft under-feathers, which are not involved in flight as are the wing and tail feathers. *(300x)*

Strands of hair from the angora rabbit may look soft and velvety, but under magnification each hair is seen to be covered with flat scales. Hair and fur are composed of keratin, the same protein material that is found in horns, hoofs, scales, feathers, and fingernails. *(700x)*

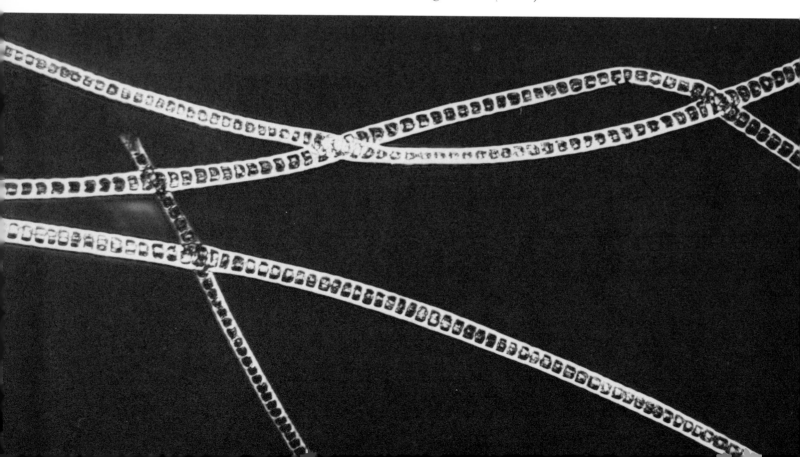

The long feelers, big compound eyes, and jointed legs of the fruit fly are very complicated and perfectly designed for finding food. Its body is covered with a stiff outside skeleton that is strong and can bend without breaking. The organs of touch and smell resemble hairs and are scattered throughout the fly's body. *(50x)*

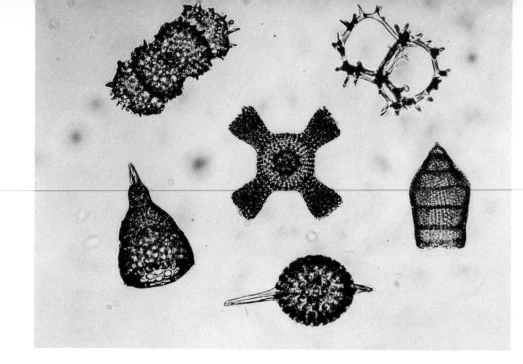

These beautiful skeletons are the coverings of sea animals called Radiolaria. When they are alive, some of these animals shine in the dark, and masses of their bodies can light up the ocean for miles around. Radiolaria shells have long been indicators of the ages of rocks. By studying the skeletons of these one-celled creatures, scientists can not only date rock sediments, but they can also determine when the sea floor itself was formed. *(275x)*

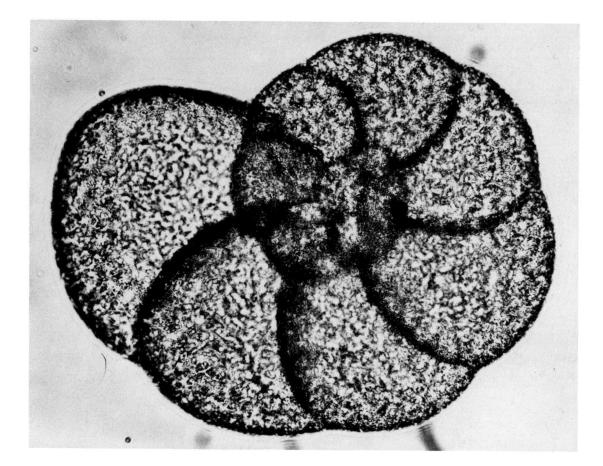

Many fish have overlapping scales, but the small tooth-like scales of the dogfish shark are scattered throughout the skin. Shown here under polarized light, they look like beautiful ceramic tiles. *(250x)*

Snails have a peculiar organ called a radula, like a long coiled tongue, which they use to scrape plants or animals from the surface of rocks. This bit of radula shows the many rows of pointed teeth that cover its surface. As a row wears out, it is replaced by the next one below it. *(350x)*

◀ Looking more like a giant sea snail than a one-celled protozoa, this Foraminifera surrounds its soft, shapeless body with a spiral shell built from a chalky material. Uncountable billions of these shells form entire reefs or mountain ranges of limestone thousands of feet thick. *(415x)*

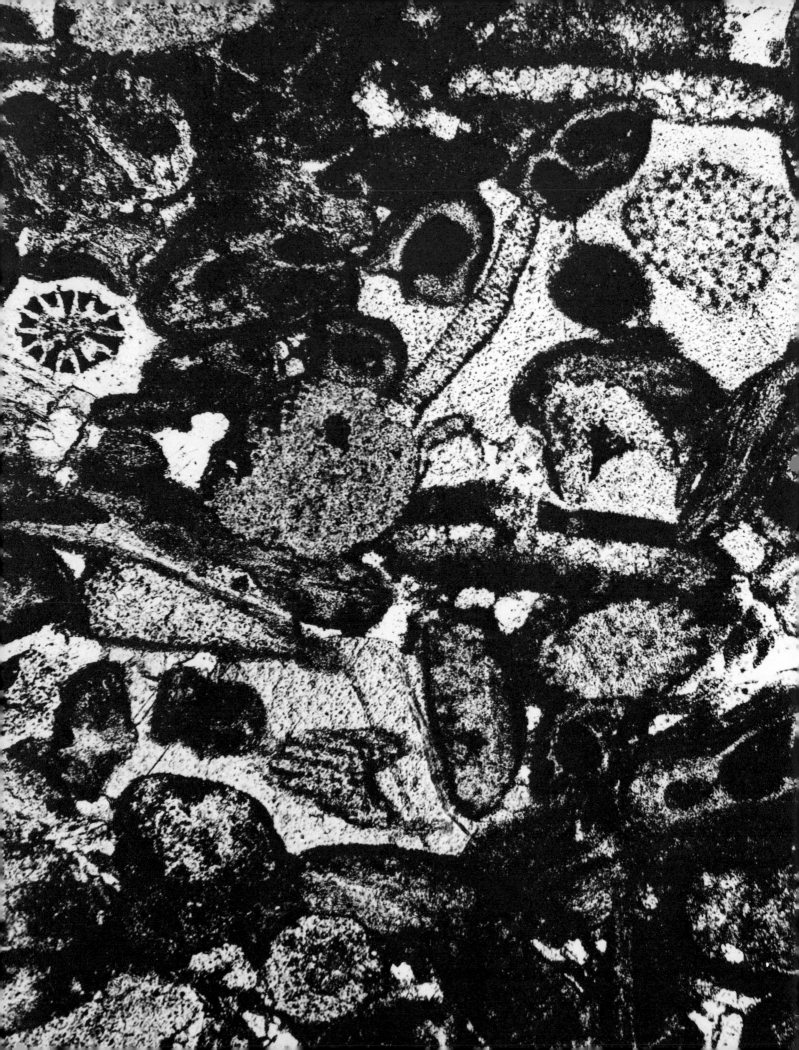

6. Crystals

WE HAVE SO FAR dealt with the living world, but under the microscope you can also look at the mineral kingdom—those things that are not alive.

Cars, transistors, aspirin, and jewelry seem totally unrelated, but they all have one important thing in common. They are all made of crystals of different shapes and sizes. In fact, glass, plastic, and some rocks are just about the only ordinary inorganic substances that are not made of crystals. Metals are actually clusters of tiny crystals pressed together into different patterns, as are most minerals. Table salt has a simple crystal shape, but most crystals have a complex and beautiful composition.

A bit of crystal forms when atoms and molecules cling together in particular patterns that repeat themselves again and again. You can watch some of the most magnificent crystals forming through the lens of the microscope. When you add a drop of water to a few grains of a salt made from one of the heavy metals like nickel or magnesium, you can see little crystals forming at the edge. Then, as the water evaporates, more crystals appear, growing faster and faster. Some grow larger in size, and suddenly, as if by magic, crystals begin to shoot out from the middle of the drop. Like a kaleidoscope they change until finally each grain of salt becomes a gallery of natural design.

◄ Not all rocks are composed of crystals. Some are made from tiny shells and bones or grains of sand, mud, or clay jammed together in layers over millions of years. These sedimentary rocks often contain fossils of plants and animals that died thousands of years ago. Scattered throughout this limestone, a sedimentary rock, you can see fossils in the shapes of wheels, spines, cones, circles, shells, and other odd bits. Scientists study fossils to learn about the evolution of the earth and its inhabitants. *(325x)*

Igneous rocks are formed when melted minerals cool and harden into crystals. In granite, a common igneous rock, combinations of feldspar, quartz, mica, and hornblende are seen as fairly large crystals or grains. Feldspar can be seen as light or white grains, the quartz as cross-hatched bits, and the mica and hornblende as the dark or black areas scattered throughout the granite. *(120x)*

When limestone is subjected to tremendous heat and pressure over millions of years, the calcite chemical it contains is changed to marble. Rocks that have been highly compressed and changed into other rocks are called metamorphic rocks.

The crystal grains in marble are rounded and crossed with minute streaks that result from impurities being forced into the limestone while it is being metamorphosed. *(110x)*

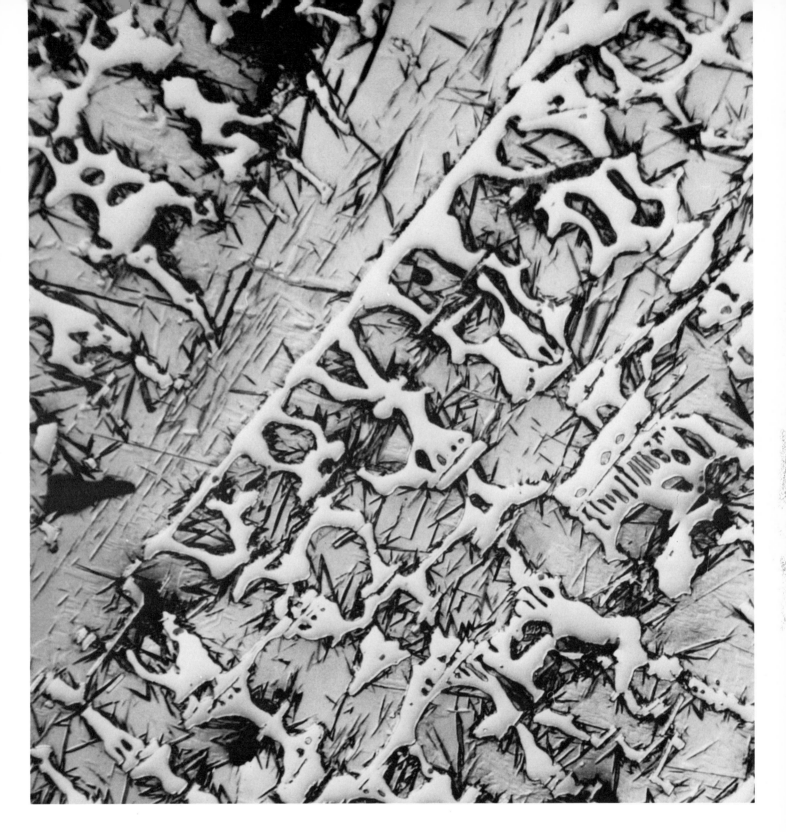

Most common industrial metals are mixtures or alloys of two or more
metals. However, some alloys like steel are composed of a metal and a
nonmetal substance like carbon. An alloy of iron and carbon, most
steel also contains small amounts of manganese, silicon, and copper,
which strengthen the alloy and make it less likely to rust.

The crystal structure of cast steel with silicon looks like snow on
ice. *(1,000x)*

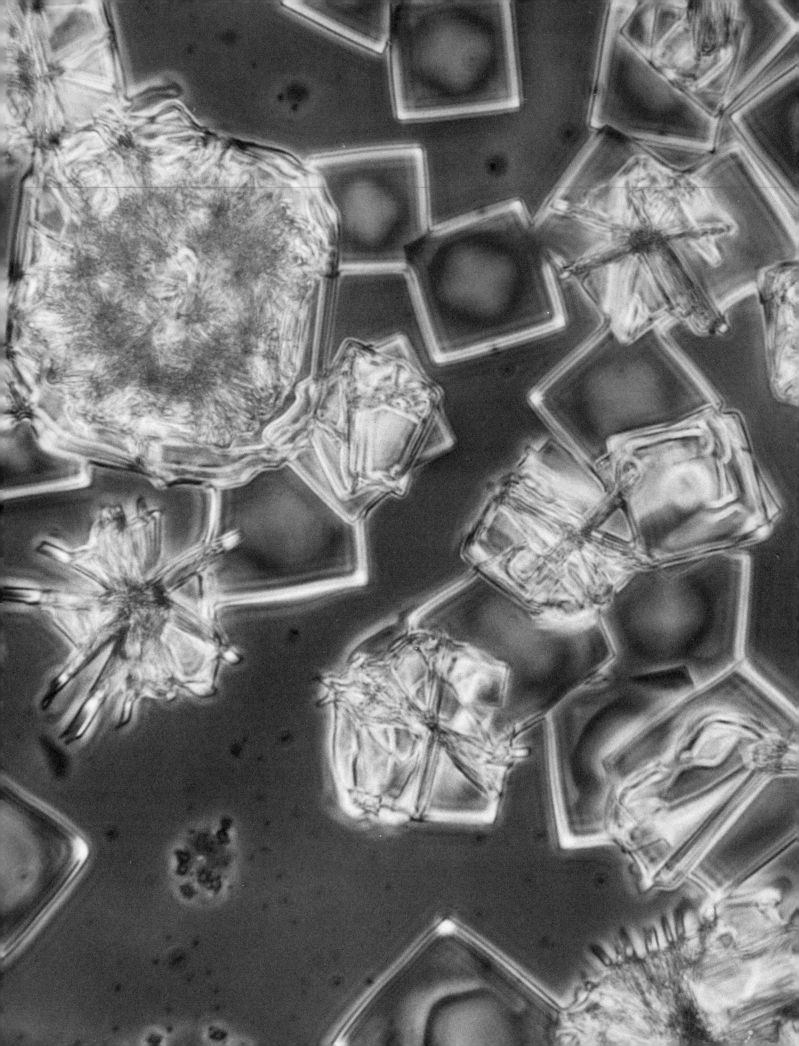

The individual crystals of magnesium sulfate appear as magnificent floral designs. Although they vary in size, each crystal has the same general shape because the arrangements of atoms within the crystals never vary. Magnesium sulfate is often used as a medicine and is known as Epsom salts. *(140x)*

◄ Some of the most unusual chemicals are derivatives of cholesterol. Although they are liquids that flow like glue, they have an inner structure of solid crystals, which, in this particular ester of cholesterol, look like blocks of ice. These "liquid crystals" are very sensitive and respond to temperature change by turning shades of red, blue, and green. *(2,000x)*

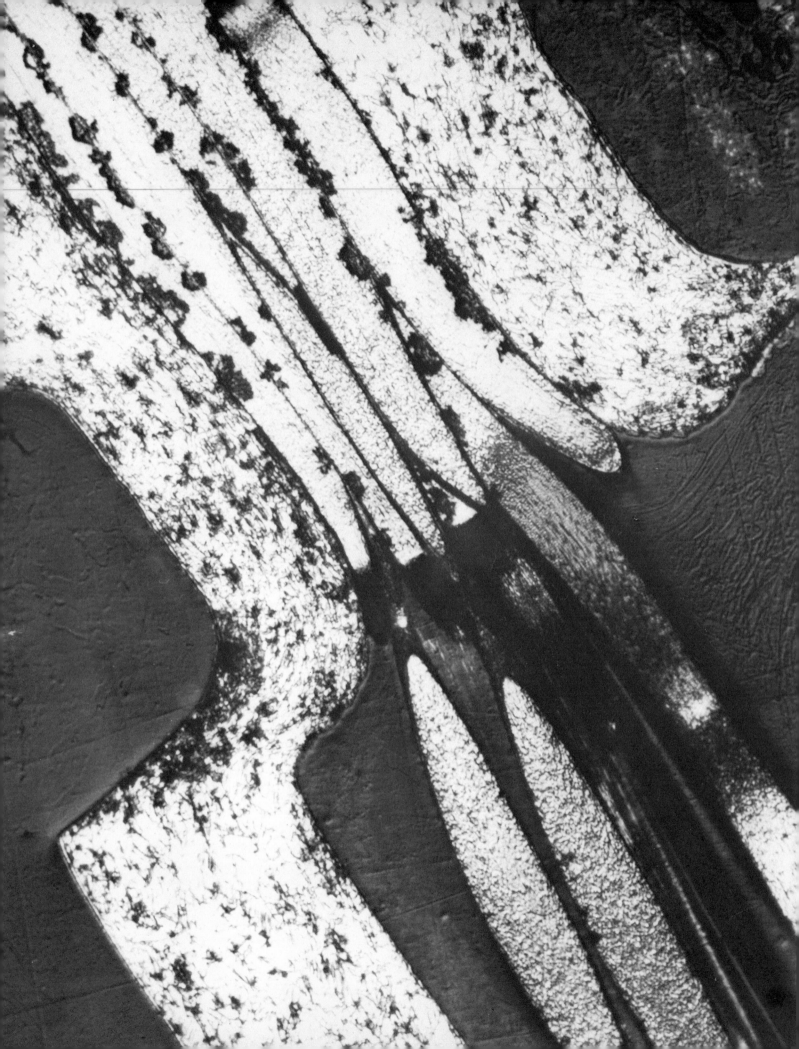

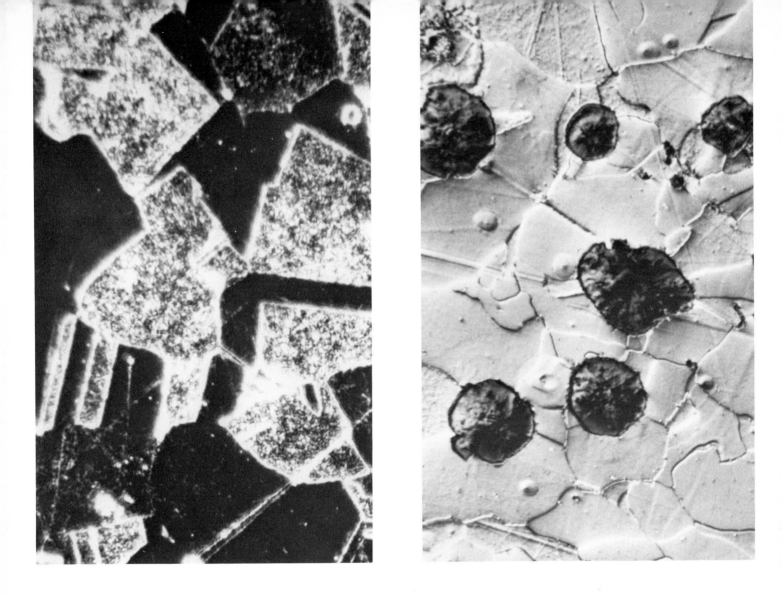

The copper and zinc crystals in brass form squares, rectangles, and triangles. Harder than copper, brass is used for many things from candlesticks to water pipes. *(3,200x)*

Cast iron is an alloy that contains varying amounts of carbon and silicon. These chemicals appear on the surface and look like morning glories in full bloom. *(1,000x)*

◄ Copper that has been shaped into terminal pins for use in electrical wiring appears as small, uneven black spots held together by tiny threads. Copper was one of the first metals to be used by man for tools, weapons, statues, and even dishes. *(1,250x)*

APPENDIX

How the Photographs in This Book Were Taken

DR. LEWIS R. WOLBERG

If you were to put a magnifying lens in front of an object, you would see a portion of that object considerably enlarged. A more powerful lens would reveal a smaller part of the object in even greater magnification. As a result, you might observe things through the lens that you could not see with the naked eye. For example, you might notice the tiny hairs on the legs of a flea that you could not otherwise see. (See title page.) There is a limit, however, to how large a magnification you can get with an ordinary magnifying lens. It is rarely possible to get more than ten or fifteen times' magnification. This is not strong enough to detect the mysteries of what exists inside things. Additionally, the usual magnifying glass does not give us a clear picture over the entire field. The center of the field may be sharp, but the edges will be blurred.

Many years ago in Holland, Anton van Leeuwenhoek invented lenses that corrected some of these defects. This was in 1673. With his more perfected lenses he was able to discover a number of microscopic animals, but even his lenses could not go beyond a certain point. Some years later, scientists took advantage of a discovery that was made even before Leeuwenhoek's time, namely that by putting a lens that curves inward on both sides (concave lens) on top of one that curves outward on both sides (convex lens), the magnification can be greatly increased. In the early days, however, lenses were so bad that the greatly magnified image was too unclear to be useful. Around 1684, Christian Huygens overcame this disadvantage by employing specially ground lenses. He put one lens system on one end of a tube and another at the opposite end. The lens near the eye was called the *eyepiece;* the one near the object was called the *objective.* This is the principle of the optical *compound microscope,* which is the instrument used for taking these pictures.

The *objective* lenses vary in the magnification they give. We can use a lens for 2½ to 100 times' magnification. The smaller the magnification, the larger the area we observe. This is helpful because sometimes we want to look at larger objects that cover an entire field with a low magnifying lens, and sometimes we want to observe tiny objects with a higher magnifying lens. The *eyepiece* lens then magnifies even more the image seen through the objective lens. For example, if the eyepiece lens gives us ten times' (10x) magnification and the objective lens give us four times' magnification (4x), the total magnification is 10 x 4 = 40 times. If the eyepiece lens is 10x power and the objective lens is 100x power, the total magnification is 10 x 10 = 1,000 times. This is about as high as we can go with an optical compound microscope, but it is powerful enough to see the smallest bacteria. The optical compound microscope is the one most frequently used to examine different types of materials in many fields, such as medicine, chemistry, geology, and metallurgy.

If we want to go beyond 1,000 times' magnification, we must use a special kind of microscope that is extremely complicated and expensive. It is called an *electron microscope,* which magnifies hundreds of thousands of times and allows us to see small viruses and large molecules.

In addition to good lens systems in the optical compound microscope, we need a powerful source of light to illuminate our object. For example, if we want to examine a piece of plant or animal that is on a slide, we place the slide on the platform of the microscope, which is called the stage, and shine a beam of light through it from below, looking at the slide through the lenses from above. A simple light system is a small concave mirror placed at an angle under the platform and pointed at the sky so that light will shine on the object. Naturally, we cannot always work in the daytime. Therefore, we must also be able to use artificial light. A small, powerful bulb—like the one used in a movie projector—is most often employed. To sharpen the light beam, we use a magnifying lens in front of the bulb. In the more expensive microscopes, there is another lens system built underneath the platform that condenses the light into a small, bright, even disc. Sometimes, when we want to look at a thick object or a piece of metal through which we cannot shine light, we have to illuminate it from above or from the side. This requires a *vertical illuminator* and lenses that shine light down a tube from above and reflect it back into the lenses.

We can use different types of cameras to make photomicrographs. The most common are 35-millimeter cameras, which have removable lenses. Once the lens is removed, the camera body is mounted over the eyepiece lens of the microscope with a special adapter. The best cameras to be used are those with a prism-focusing device and a ground glass. The image on the ground glass can be seen easily, and we can move the object around on the stage until we can see exactly the object and composition we want to photograph. Focusing is accomplished by adjusting the focusing knob on the microscope. If the camera has a built-in light meter, exposure is easily accomplished. Otherwise, you have to experiment with a light meter placed against the ground glass. The principle is the same if larger cameras are used. Cameras used in commercial photomicrography are sometimes 2¼" x 2¼", 2¼" x 2¾", or 4" x 5". Because the cameras are heavier, they are mounted on a separate stand.

In order to bring out contrasts and colors in the specimens we are photographing, we often use *polarizing filters*. One filter is placed in the light beam before it reaches the object specimen, and one is placed over the specimen. When one of the polarizing filters is rotated, interesting and dramatic effects can be obtained, particularly when we look at rocks, crystals, and certain materials and tissues. As we rotate the polarizer, we notice changes of color from violet to indigo, to blue, to green, to yellow, to orange, to red. This is because the polarized white light is being split into its different component colors by the crystals, in the same way as bits of moisture in the air break up the sunlight into a rainbow. The same area we are observing can be changed into colors of our selection and will photograph as such with color film. When we use black and white film, we can control the shading of light and dark areas.

In figuring out the total magnification in the final picture, it is necessary to multiply the magnification of the microscope by the degree of the photo enlargement. For instance, if we use an object of 4x power and an eyepiece of 8x power, we will obtain a negative that has a picture of 32x magnification. If we enlarge the negative in our final print

10 times, the total enlargement is 32 x 10 = 320. The objects printed or projected on the paper or screen are, therefore, 320 times larger than those on the original slide.

The technique of slide preparation is easily learned. It is a process of hardening, staining, placing in a paraffin block, shaving with a slicing tool *(microtome),* and mounting the shaving on a thin glass slip. This preparation is frequently not necessary, however, because a variety of excellent prepared slides can be purchased from biological supply houses. To observe many crystal formations, you can dissolve chemicals in hot water and place a drop on a glass slide. As the water evaporates, the dried-out crystals are easily photographed. Thin whole objects like tiny insects need not be sectioned, but can simply be placed on a glass slide. You can also spend hours observing and photographing worlds of tiny creatures in a single drop of water.

The actual instrument used in the pictures shown here was a Zeiss Research Microscope equipped with plano-apochromatic objective lenses and a vertical illuminator and epiplan lenses for opaque objects. The camera was a Rolleiflex SL 66, an excellent 2¼ x 2¼" prism-focusing camera. The exposure meter was a Gosser Lunasix.

BIBLIOGRAPHY

Allen, R. M. "Photomicrography," in *The Encyclopedia of Photography*. New York: Greystone Press, 1967, pp. 2823-38.

Croy, O. R. *Creative Photomicrography*. New York: Amphoto, 1968.

Johnson, G. *Hunting with the Microscope*. New York: Sentinel Books, 1963.

Wolberg, Lewis R. *Micro-Art: Art Images in a Hidden World*. New York: Harry N. Abrams, Inc.